The Why, How, and What of Existence

Through the Symbolism of the Wheel

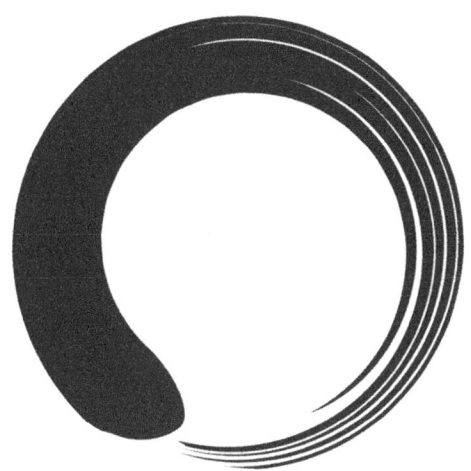

Copyright © 2022 Vlad Korbel

All rights reserved. No part of this book, in part or in whole, may be reproduced, transmitted, or utilized, in any form or by any means, electronic or mechanical, including photocopying, recording, or by any information storage and retrieval system, without permission in writing from the publisher, except for brief quotations in critical articles, books and reviews.

Disclaimer: Opinions expressed in this book do not represent Novartis, but solely personal views of Vlad Korbel, hence cannot be associated with Novartis in any way. References are made to previously published content at Stanford University d.school expert talk in 2021 and at the Design Thinking Summit in Berlin in 2019, which Novartis both approved.

ISBN 13: 978-1-56184-045-8
ISBN 10: 1-56184-045-9

New Falcon Publications First Edition 2022

The paper used in this publication meets the minimum requirements
of the American National Standard for Permanence of
Paper for Printed Library Materials Z39.48-1984

Printed in USA

NEW FALCON PUBLICATIONS
2046 Hillhurst Avenue
Los Angeles, CA 90027
www.newfalcon.com
email: info@newfalcon.com

The Why, How, and What of Existence

Through the Symbolism of the Wheel

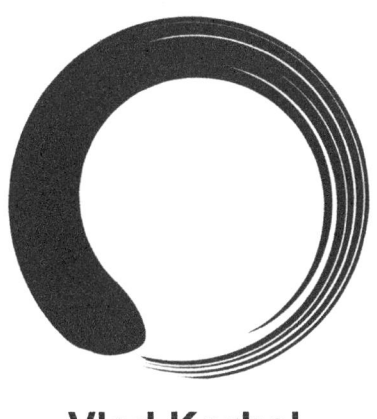

Vlad Korbel

NEW FALCON PUBLICATIONS
Los Angeles, California, U.S.A.

"In the beginner's mind there are many possibilities, but in the expert's there are few."

"When your mind is compassionate, it is boundless."

– Shunryu Suzuki

*This book is dedicated to my family
for the much-needed
love.*

*To the great people in drug development who work hard
to help their fellow human beings; and the Stanford d.school,
which has taught me the right design (thinking) mindset, methods,
and tools, I've been using throughout my discoveries.*

*Big thanks to Alan Watts, Albert Hofmann, Stan Grof,
Terence McKenna, Ram Dass, and many other
inspirational people whose thoughts and spirit
guided me through this journey.*

Introduction

This book summarizes my experiences during the years I was researching human behavior in the drug development ecosystem and studying at the Stanford Innovation Program.

Extreme curiosity and the strongest will to help this world led me to these revelations through multiple journeys of inquiry to the Supreme Self.

VLAD KORBEL

Why?

I myself, have undergone a journey of inquiry to the Supreme Mind, which one might, for the sake of simplicity, call God. Here I must state that by God, I mean the supreme intelligence, the creator, sustainer, and destroyer. I don't mean any of those prophets of the past who were mistakenly worshiped.

—

My first inquiry to Him/Her/It was:

Me: "Please, I need your help, the people ruling this world are about to destroy our planet."

He/She/It: "There is nothing I can give you. You already have it."

Me: "So what is it?"

He/She/It: "It's The Love; it's the only one which is worth following in life. There is nothing else. There is nothing like that one. It's the only one we have. Without Love people wouldn't do anything."

He/She/It: "Would you do this if I give you that?"

Me: "What do you mean by 'that'?"

He/She/It: "This, I can give you anything you want."

Me: "No, I don't need it. I won't do it."

He/She/It: "So you see. Now I need your help. Can you help me?"

Me: "I have nothing to offer; the only thing I have is my life, but I'm not sure it will help. I can offer it to you."

He/She/It: "So do you want to die? I can do it right now."

Me: "No, please, I don't want to die."

After that, I was sent to other worlds, where I met beings inhabiting those places. One of the places I could call Heaven, Origin, or simply the "Other side" while another felt more like Hell.

After some weeks, I was still very curious, so I went on a second journey of inquiry. I was a little afraid to ask again, but my curiosity about that was stronger than my fear.

So I asked again: So what is it?

He/She/It: "I don't know. Nobody really knows."

Me: "I want to know!"

He/She/It: "Watch this!"

Me: I have no words to describe it, but the nearest I can get is "Sat Chit Ananda*".

The ultimate truth - consciousness - bliss.

He/She/It: "This is it, and you are it!
(Tat Tvam Asi**)."

* *Sat Chit Ananda (Sanskrit) = truth, consciousness, bliss = when the Atman (the self) experiences the Brahman (the Self) as one.*

** *Tat Tvam Asi (Sanskrit) = You (the self) are Brahman (The Self).*

My takeaway from this experience is: We are all part of one consciousness, which creates, sustains, and destroys us. We are all one family both on Earth and in the Universe. We have nothing to fear of when we die. We do not eternally disappear but remain in the consciousness.

We can have a civilized society where everyone gets what they really need and not necessarily accumulate what they think they need, while taking it away from their brothers and sisters.

Om Namah Shivaya

All consciousness is one

The Why, How, and What of Existence

VLAD KORBEL

How?

In *design*, or *design thinking*, creativity and imagination are where the "pedal meets the metal" or the "rubber meets the road". If we can imagine it, it can be done. This is one of the key qualities that *design* can offer to humanity. Based on *user needs* researched during problem discoveries, the 'why', we come up with a design brief and ideate multiple possible futures of the 'what', and 'how' are we going to get there. Once we decide on the best one, we make it happen. In this section, I'm going to talk about the 'what' and 'how'. What is the goal or vision and how to get there.

For designers and innovators, it is important to be able to think big in order to introduce a change. The bigger the design gets, the bigger the change it can be.

That we are not God is the fundamental assumption–the unquestionable dogma, which our modern world operates upon, and according to the Catholic Church, questioning it would be a heresy.

On the contrary, entities like the Hermes in the Greek mythology or Thoth in Egyptian mythology (Mercury in the age of the Roman Empire)–all embodying the archetype of wisdom, taught us a lesson. The lesson they taught us is that if we want to understand God, we have to start thinking as equal to God.

Hermes, Thoth, Mercury who are believed to be the same entity, sent us a message in one of the most destroyed scriptures throughout the Middle Ages.

From the book Hermetica, which one part of it, 'To Asclepius', reads as follows: "If then you do not make yourself equal to God, you cannot apprehend God; for like is known by like.

Leap clear of all that is corporeal, and make yourself grown to a like expanse with that greatness which is beyond all measure; rise above all time and become eternal; then you will apprehend God. Think that for you too nothing is impossible; deem that you too are immortal, and that you are able to grasp all things in your thought, to know every craft and science; find your home in the haunts of every living creature; make yourself higher than all heights and lower than all depths; bring together in yourself all opposites of quality, heat and cold, dryness and fluidity;

think that you are everywhere at once, on land, at sea, in heaven; think that you are not yet begotten, that you are in the womb, that you are young, that you are old, that you have died, that you are in the world beyond the grave; grasp in your thought all of this at once, all times and places, all substances and qualities and magnitudes together; then you can apprehend God."

— *Hermetica: The Greek Corpus Hermeticum and the Latin Asclepius by Hermes Trismegistus; page 41.*

In these ancient times, it had already been understood that there are no limits—imagination is infinite. The only thing we must do is to truly believe these impossible dreams, in order to become like God. But what we have to understand as well is the other limit of the spectrum of the wave—the emptiness, zero or void. To be able to explore both sides of the spectrum. We have to understand this humbling principle—to know that we know nothing. How else would we be able to grasp the infinite or the absolute if we do not understand the emptiness? As Socrates taught us, the true ***wisdom*** lies in this understanding.

Thus, it is the understanding of both limitations of our thinking to the extreme, which then becomes the vehicle for us being able to become Gods in its fundamental sense.

In Zen, one of the greatest riddles is the path to enlightenment by passing through the gate-less gate. By the physical means "No-one can pass through the gate-less gate."

So in order to be able to pass through the gate-less gate, we have to become 'no-one'.

By admitting to ourselves that we do not know, suppressing the ego of the expert and being truly compassionate, we are then able to find the right problem to focus on. Then, by imagining the unimaginable, find the ideal future and how to get there.

Without these qualities, design will remain small, and the positive change, which design can bring about, will remain marginal.

The Why, How, and What of Existence

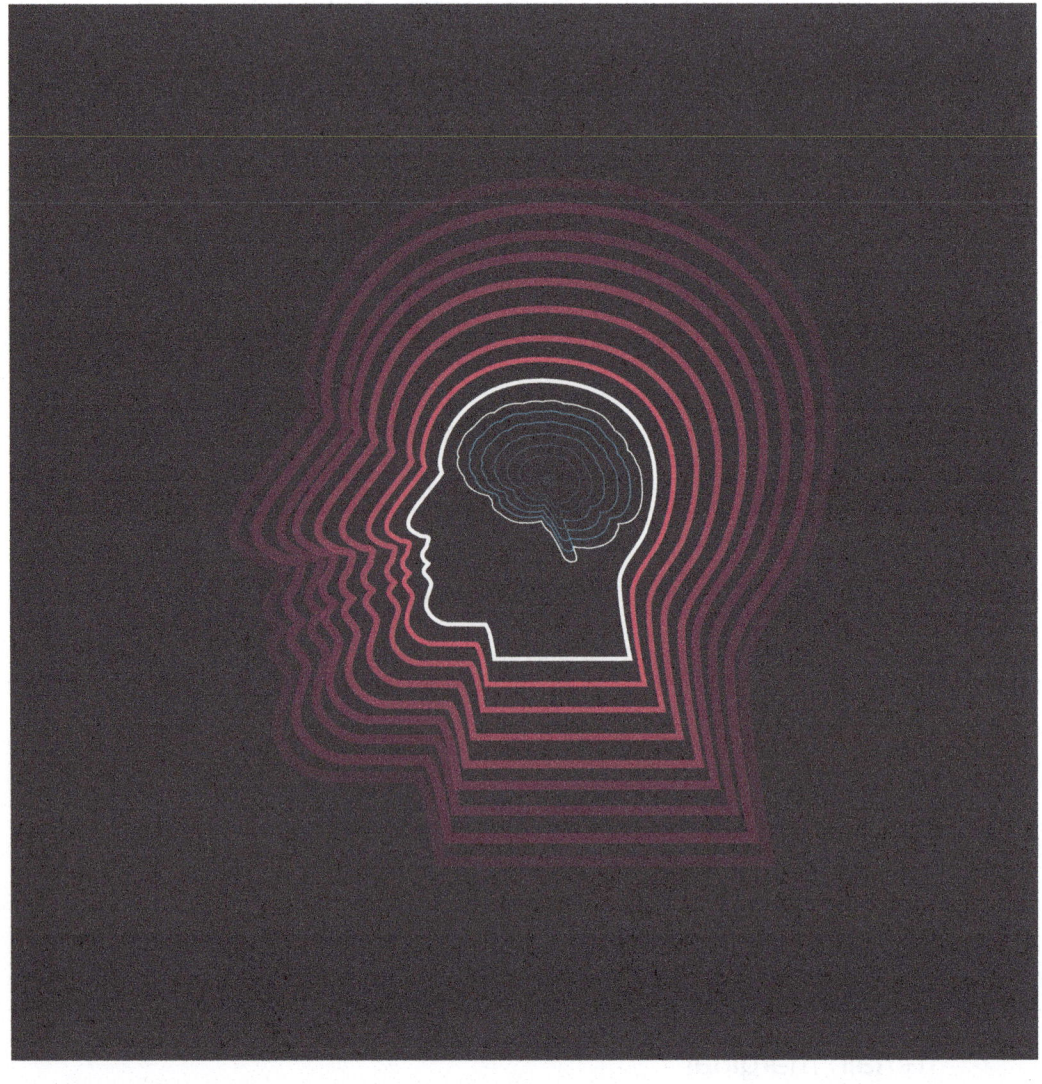

What?

These verses have been inspired by Sri Ramana Maharshi. Ramana Maharshi was a Hindu sage and a liberated being who lived in India in the first half of the 20th century and who invented a method of inquiry into one's true nature based on the question "Who am I" and on the "Neti Neti" (Not this, not that) principle of negation described in the Upanishads and advocated by Adi Shankara, one of the foremost Advaita (non-dualist) philosophers of India.

Who am I?

My name, title, and identity.
— I am not.

My body, environment, and my possessions.
— I am not.

My experiences, achievements, actions, needs,
and desires.
— I am not.

My mind: that which thinks that I am not;
My soul which feels that I am not,
My destiny which determines that I am not.
— I am not.

The ever-lasting, omnipresent existence, consciousness, bliss;

The Spirit beyond duality and oneness;

The no-one, yet the absolute;

– I am.

After He's taken me through Heaven and Hell, He manifested Himself as a ray of light of exact straightness;

I felt being one with that ray of light as if it was the only thing which truly existed;

He let me be Him for a little time, flying through space in all directions I wanted;

He shined through my mind, paused, turned around, pointed at Himself and asked:

"Who am I?"

He laughed. I laughed too, bathing in love and tears, knowing that we are one.

The Why, How, and What of Existence

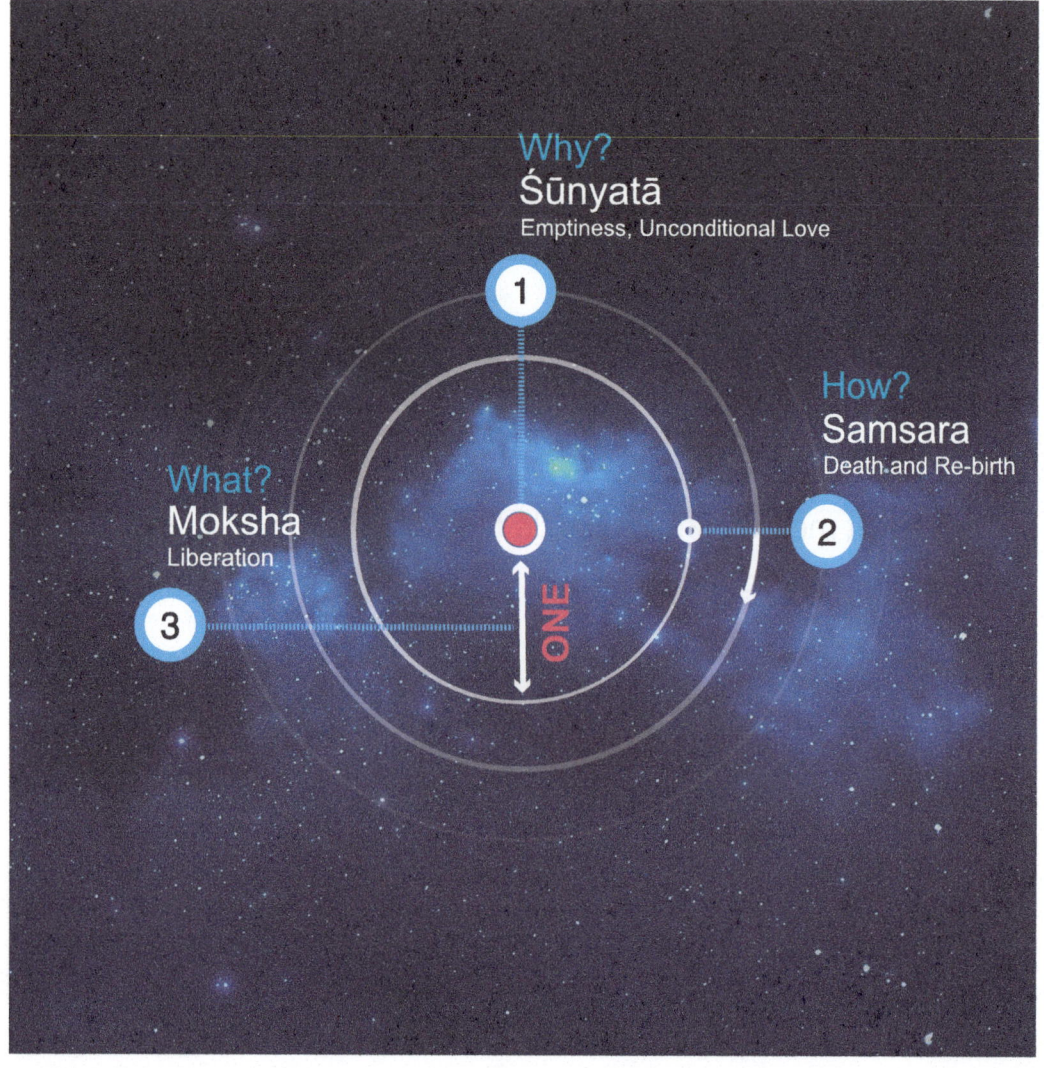

The Why, How, and What of Existence

"Love never fails. But if there are prophecies, they shall fail; if there are tongues, they shall cease; and if there is knowledge, it shall vanish. For we know in part, and we prophesy in part. But when that which is perfect comes, then that which is imperfect shall pass away. So now abide faith, hope, and love, these three. But the greatest of these is love."

– Paul 1:13

"There is an extremely powerful force that, so far, science has not found a formal explanation to. It is a force that includes and governs all others, and is even behind any phenomenon operating in the universe and has not yet been identified by us. This universal force is LOVE."

– Albert Einstein

"God seeks comrades and claims love, The devil seeks slaves and claims obedience."

– Rabindranath Tagore

"The last judgment is the witness examined by your higher self—how you applied the law of LOVE."

– Peter Dunoff

Vlad Korbel

is a designer, researcher, occasional speaker, and philosopher.

VLAD KORBEL

He was born in Prague, the city of the ancient alchemists, in 1983. After the Velvet revolution, he was inclined toward the rave counterculture, which Terence McKenna has philosophically inspired. He received a master's degree in economics in 2010. Vlad currently lives in Prague, while his second home is in Quito, Ecuador. He also spent significant time in Basel, Switzerland, researching human behavior at Novartis, the successor company of Sandoz, where Albert Hofmann made his famous discovery in 1938. Vlad studied the Stanford Innovation program, which he graduated in 2019, and used this knowledge for his research in the drug development ecosystem. He became fascinated by some of the fundamental Hindu schools of thought, such as Advaita Vedanta and Shaivism, around the same time. These concepts inspired the design of a clinical trial control center, SENSE, which he designed for Novartis. Multiple world-renowned media outlets have published this design, thanks to the embodiment of the wheel's symbolism—using a metaphor for the fundamental behavioral reasoning—love. He has undergone several journeys of inquiry to the 'Supreme Self', based on which he wrote this spiritual book.

Some Other Titles From New Falcon Publications

Aha! The Sevenfold Mystery of the Ineffable Love	–Aleister Crowley
An Insider's Guide to Robert Anton Wilson	–Eric Wagner
Bio-Etheric Healing	–Trudy Lanitis
Undoing Yourself With Energized Meditation and Other Devices, Secrets of Western Tantra: The Sexuality of the Middle Path, Dogma Daze	–Christopher S. Hyatt, Ph.D.
Rebels & Devils; The Psychology of Liberation–	Edited by Christopher S. Hyatt, Ph.D.
Aleister Crowley's Illustrated Goetia, Sex Magic, Tantra & Tarot: The Way of the Secret Lover, Taboo: Sex, Religion & Magick	–C. Hyatt, Ph.D., and Lon Milo DuQuette
Pacts With The Devil, Urban Voodoo: A Beginner's Guide to Afro-Caribbean Magic	–Jason Black and Christopher S. Hyatt, Ph.D.
The Psychopath's Bible	–Christopher S. Hyatt, Ph.D., and Jack Willis
Ask Baba Lon	–Lon Milo DuQuette
Aleister Crowley and the Treasure House of Images	–J.F.C. Fuller, Aleister Crowley, Lon Milo DuQuette and Nancy Wasserman
Enochian Sex Magic and How To Workbook	–Aleister Crowley, Lon Milo DuQuette and Christopher S. Hyatt, Ph.D.
Enochian World of Aleister Crowley	– Lon Milo DuQuette and Aleister Crowley
Info-Psychology, Neuropolitique, The Game of Life, What Does WoMan Want?	–Timothy Leary, Ph.D.
Nonlocal Nature: The Eight Circuits of Consciousness	–James A. Heffernan
on What is	–Ja Wallin
Rebellion, Revolution and Religiousness	–Osho
Reichian Therapy: A Practical Guide for Home Use	–Dr. Jack Willis
Shaping Formless Fire, Seizing Power, Taking Power, The Magick in the Music and Other Essays	–Stephen Mace
The Illuminati Conspiracy: The Sapiens System	–Donald Holmes, M.D.
The Secret Inner Order Rituals of the Golden Dawn	–Pat Zalewski
The Why, Who, and What of Existence	Vlad Korbel
Steamo Goes to Havana, The Social Epidemic of Child Abuse	Michael Miller, M.Ed., M.S., Ph.D.
Woman's Orgasm: A Guide to Sexual Satisfaction	–Benjamin Graber, M.D., and Georgia Kline-Graber, R.N.

Other Titles by J. Marvin Spiegelman, Ph.D.

A Modern Jew in Search of Soul
Buddhism and Jungian Psychology
Catholicism and Jungian Psychology
Hinduism and Jungian Psychology
Mysticism, Psychology and Oedipus - A Small Gem
Protestanism and Jungian Psychology
Psychotherapy and Religion at the Millennium and Beyond
Psychotherapy as a Mutual Process
Reich, Jung, Regardie & Me - The Unhealed Healer
Rider, Haggard, Henry Miller & I - The Unpublished Writer
Sufism, Islam and Jungian Psychology
The Knight - A Small Gem
The Nymphomaniac
The Quest - Further Adventures in the Unconscious
The Tree of Life - Paths in Jungian Individuation
The Wisdom of J. Marvin Spiegelman Vol. I - Selected Writings
The Wisdom of J. Marvin Spiegelman Vol. II - Psychology and Religion

Other Titles by Dr. Israel Regardie

A Garden of Pomegranates
A Practical Guide to Geomantic Divination - A Small Gem
Attract and Use Healing Energy - A Small Gem
Be Yourself - A Guide to Relaxation and Health
Ceremonial Magic
Dr. Israel Regardie's Definitive Work on Aleister Crowley, The Eye In The Triangle
Healing Energy, Prayer and Relaxation
How To Make and Use Talismans - A Small Gem
Israel Regardie's The Foundations of Practical Magick
My Rosicrucian Adventure
Mysticism, Psychology and Oedipus - A Small Gem
Practical Magick - A Small Gem
Teachers of Fulfillment
The Art and Meaning of Magic - A Small Gem
The Body-Mind Connection, A Path to Well-Being - A Small Gem
The Complete Golden Dawn System of Magic
The Complete Golden Dawn System of Magic Book 1 - Ltd. Edition
The Complete Golden Dawn System of Magic Book 2 - Ltd. Edition
The Complete Golden Dawn System of Magic - The Black Edition
The Eye in the Triangle: An Interpretation of Aleister Crowley
The Golden Dawn Audio CDs, Vol. 1, Vol. 2, and Vol. 3
The Legend of Aleister Crowley
The Magic of Israel Regardie
The Middle Pillar
The Philosopher's Stone
The Portable Complete Golden Dawn System of Magic
The Tree of Life
The Wisdom of Israel Regardie - Vol. I: Selected Introductions, Prefaces and Forewords
The Wisdom of Israel Regardie - Vol. II: Selected Essays and Commentaries
The Wisdom of Israel Regardie - Vol. III: Selected Articles, Introductions, Prefaces and Forewords
What You Should Know About the Golden Dawn
Wilhelm Reich, His Theory And Techniques
Aha! (Dr. Israel Regardie and Aleister Crowley)
Roll Away The Stone/The Herb Dangerous (Dr. Israel Regardie and Aleister Crowley)

 **NEW FALCON PUBLICATIONS**

At the Falcon website you can:

- Browse the online catalog of all our great titles, including books by Robert Anton Wilson, Christopher S. Hyatt, Israel Regardie, Aleister Crowley, Timothy Leary, Osho, Lon Milo DuQuette and many more
- Find out what's available and what's out of stock
- Get special discounts
- Order our titles through our secure online server
- Find products not available anywhere else including:
    - One of a kind and limited availability products
    - Special packages
    - Special pricing
- And much, much more

MANY OF OUR TITLES AVAILABLE ON KINDLE!
Please visit our website at http://www.newfalcon.com

The Why, How, and What of Existence

Through the Symbolism of the Wheel

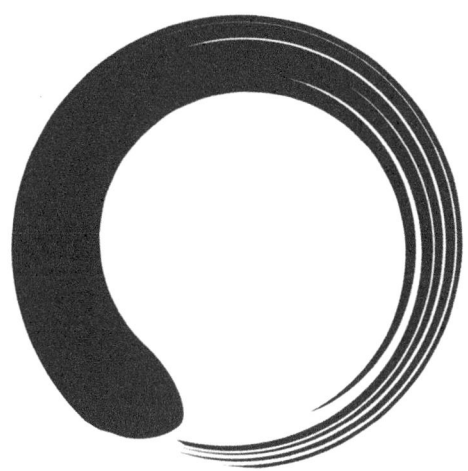

Copyright © 2022 Vlad Korbel

All rights reserved. No part of this book, in part or in whole, may be reproduced, transmitted, or utilized, in any form or by any means, electronic or mechanical, including photocopying, recording, or by any information storage and retrieval system, without permission in writing from the publisher, except for brief quotations in critical articles, books and reviews.

Disclaimer: Opinions expressed in this book do not represent Novartis, but solely personal views of Vlad Korbel, hence cannot be associated with Novartis in any way. References are made to previously published content at Stanford University d.school expert talk in 2021 and at the Design Thinking Summit in Berlin in 2019, which Novartis both approved.

ISBN 13: 978-1-56184-045-8
ISBN 10: 1-56184-045-9

New Falcon Publications First Edition 2022

The paper used in this publication meets the minimum requirements
of the American National Standard for Permanence of
Paper for Printed Library Materials Z39.48-1984

Printed in USA

NEW FALCON PUBLICATIONS
2046 Hillhurst Avenue
Los Angeles, CA 90027
www.newfalcon.com
email: info@newfalcon.com